Quantum Physics for Beginners Who Flunked Math and Science

Quantum Mechanics and Physics Made Easy Guide in Plain Simple English

Donald B. Grey

I0480476

1

Quantum Physics for Beginners

Bluesource And Friends

This book is brought to you by Bluesource And Friends, a happy book publishing company.

Our motto is **"Happiness Within Pages"**

We promise to deliver amazing value to readers with our books.

We also appreciate honest book reviews from our readers.

Connect with us on our Facebook page www.facebook.com/bluesourceandfriends and stay tuned to our latest book promotions and free giveaways.

Quantum Physics for Beginners

Quantum Physics for Beginners

Table of Contents

Quantum Physics for Beginners

Quantum Physics for Beginners

Introduction

'Look again at that dot. That's here. That's home. That's us. On it everyone you love, everyone you know, everyone you ever heard of, every human being who ever was, lived out their lives. The aggregate of our joy and suffering, thousands of confident religions, ideologies, and economic doctrines, every hunter and forager, every hero and coward, every creator and destroyer of civilization, every king and peasant, every young couple in love, every mother and father, hopeful child, inventor and explorer, every teacher of morals, every corrupt politician, every "superstar," every "supreme leader," every saint and sinner in

the history of our species lived there—on a mote of dust suspended in a sunbeam." (Sagan, 1994).

The pale blue dot has always been our home, but we have never understood it, its place in the Universe, or the magic that makes it tick—not as well as we do today.

Until the beginning of the 20th century, we had managed to be quite smart about our discoveries. We understood the order of things in the Universe—an order that made sense and to which we could relate to.

And then, somewhat out of the blue, came a theory that was wilder than our imagination, crazier than our most peculiar thoughts, and more misunderstood than the beginning of life in the Universe: Quantum mechanics.

As if to show us all that we're not the kings of our Universe and that we might, after all, know nothing about everything that surrounds us, quantum theory lay down a long list of questions and told us: *Guys, I can help you answer these, but you will first have to solve the riddle.*

Arrogant as we usually are, we thought we had it all figured out. We learned that the small world was nothing like the big one and that at the most granular levels of the Universe, strangeness was at home. Gone was the logic and the order of things in the Universe, which we took so much pride in just a few years before. Gone was the peaceful sleep of a bunch of bright minds that held our hands throughout the beginning of the 20th century.

Culturally, socially, politically, economically, and even from an artistic point of view, we were making somewhat of a quantum leap into a new world. Bear in mind here: "Quantum leap" is to be understood in its metaphorical sense, not the scientific one we will discuss later on in the book.

Scientifically, we are jumping into our future without even realizing what is going on, where we are going, and how we came to "wake up" this way.

Thanks to a handful of brilliant minds in physics, this jump was much smoother than it would have been otherwise—because what quantum theory revealed was unsettling and weird as if to laugh in our face for ever having thought we had it all figured out.

Today, about one century later, from the moment the quantum theory was first formulated, we still don't know much about it. We have the groundwork done by names like Niels Bohr, Erwin Schrodinger, Albert

Quantum Physics for Beginners

Einstein, and many others that have come since them. But we still lack the fundamental information that would finally allow us to say: Yes, we got it *right* this time around.

Quantum physics is one of the oddest and most fascinating branches of science. Somewhat at the confluence between pure science (it doesn't get more "science-y" than physics, right?) and mysterious spirituality, quantum mechanics is poorly understood across the entire board.

I'd be willing to bet that a large percentage of our physics teachers in school did not fully understand the basic tenets of quantum theory, as little as it may be included in school curricula everywhere in the world.

That is a real pity. As I have come to learn, perhaps too late in life, quantum physics is eternally fascinating—a lot more exciting than the deterministic equations we were being taught as sixth or seventh-grade students. It's most definitely a lot more intriguing than empty definitions that do not correlate with *realness* and *palpableness* of physics as a whole.

Even more, if we had been presented with the fascinating tidbits of quantum physics back in school,

Quantum Physics for Beginners

we'd all be much more interested in science and physics. Whatever it is that the smart guys are doing in the Universities of the world right now would be more interesting for all of us.

It's never late to start, though. The educational system has apparent gaps in it, but that does not mean you have to live the rest of your life with these gaps in your education and knowledge repository.

Thankfully, we have the Internet at our fingertips, and we can gain easy access to a lot more information than our parents did back in the day, for example. We can select the kind of information we feed our minds with, and can choose the type of information we feed the minds of our children with, too.

We have access to thousands upon thousands of books available for purchase and download in a matter of seconds—something we would never have thought was possible, but yet has grown to be very much a part of our daily lives.

The book you are holding in front of your eyes is proof that the impossible and improbable can sometimes become tangible reality. It may also be beneficial that we keep searching for answers, and quantum physics might be *where* we will find them.

Quantum Physics for Beginners

I wrote this book with one main goal in mind: To help complete beginners in quantum mechanics understand its most basic tenets and most amazingly fascinating ideas. If you are here reading this introduction, it probably means you are curious to learn more about this field, and I cannot help but congratulate you on your choice.

When I started writing this, I wanted to make sure I pulled together only those bits of information that may actually be of interest to most non-experts. As such, I brought science-based info that makes one wonder, ask the right questions, and helps us understand the unusual and beautifully-messy nature of our reality.

I wanted to make this easy for you, so I chose not to go in-depth on the topics approached in this book. I also made a conscious choice not to present you with equations or formulas, as I know just how frightening those can be.

Instead, I attempt to explain the complex and sometimes downright confusing basic elements of quantum mechanics so that you can easily understand it. Even more than that, I try to describe them in an interesting way to make it exciting even for someone who never had any interest in science—or physics in particular (because I promise you, it will be).

Quantum Physics for Beginners

We will begin where all of this started in the first place: With waves and particles. If, in the deterministic view of the world proposed by Sir Isaac Newton, things were pretty clear and light was a wave, while pretty much everything else was perceived as particles, quantum theory came to dismember this view of the world and propose a new, enhanced, and a lot more ambiguous one: What if we thought of *everything* as being both a wave and a particle?

Quantum theory starts there, and it evolves to the point of being on the fringe with science-fiction. From waves, particles and the different interpretations of the original quantum theory to identical particles and the oddity of the laws of quantum physics, humanity has leapt forward in an attempt to understand what we are *really* made of.

We (and absolutely everything in the Universe) are made out of very, very, small pieces. Some of these smallest pieces come under the name of "neutrino," and they will make the main focus of our third chapter in this book. You might think "small particles" are not that interesting, but when you get acquainted with neutrinos, I promise that you will change your opinion.

Quantum Physics for Beginners

Further on, we will discuss some of the essential rules of quantum mechanics: Quantum leaps, quantum entanglement, quantum spins and, in the end, how even scientists had to admit that we have to live with a certain degree of uncertainty in our lives, even when it comes to something as precise and mathematical as physics itself.

In the fifth chapter of this book, we will talk about one of the most debated and fascinating topics connected to quantum physics: The uncertainty principle, as elaborated by Werner Heisenberg. Believe it or not, the first couple of decades since the birth of quantum mechanics were filled with scandal. You might not think of physicists as very "scandalous" people in general. However, the debates they had around the uncertainty principle and everything connected to the fundamentals of quantum physics will prove you wrong.

To give you a bit of a spoiler alert (just to spike your curiosity), I will mention that Albert Einstein, perhaps the single most famous physicist in the world (even to date), was on the "wrong" side of things when it comes to quantum physics. You'd think that the name that is on everyone's lips when thinking of physics as a whole would have contributed to the birth of the physics branch that shaped everything from the 20th century onwards. However, his involvement with the quantum world was a lot more complicated than most people think.

The sixth chapter of our book will be about quantum fields theory, a postulation that is outrageously fascinating and which might alter the way we see the Universe itself. In a world where quantum physics has been proven right, time and again, the quantum fields theory came as a complement to make the field richer, more resilient, and reliable. As such, it is a topic we do not want to skip at all.

Next up, we will move to Erwin Schrodinger and his cat—perhaps one of the most famous cats in the world of science (of equal fame and reputation as Pavlov's dog in the world of psychiatry). Believe it or not, the story of Schrodinger's cat lies at the core of the scandal between classical physics and quantum physics, so it is a story with which you should get acquainted.

In the eighth chapter of our book, we will discuss what might be one of the most fantastic topics related to quantum physics: Teleportation. I don't want to spoil it for you, but I promise that the information you will gain in this chapter (and from the entire book, actually) will make humankind's wildest dreams seem more feasible than ever.

Quantum Physics for Beginners

Chapters nine and 10 will be all about the Zeeman Effect and a theorem that might go against the idea presented in the eighth chapter. I have included it in the book because I believe it is important for you to have the full picture of these matters and become aware of where we stand in our evolution as a species and our way ahead.

Chapter 11 of this book will be dedicated entirely to Bell's theorem—the one finding that finally put the war between Einstein and quantum physicists to rest. Furthermore, in the last chapter, we will talk more about Albert Einstein, his complicated relationship to quantum mechanics, and why nobody should dismiss him as a scientist (even though his opinion on quantum physics was not always correct).

By the end of this book, I hope that you will have a fuller, broader picture of what quantum physics is, why it is so important for our evolution from hereon, and what it is that scientists are working on right now. (Spoiler alert number two: It's a reconciliation that has been waiting for nearly one hundred years to happen).

Whether we like it or not, our quantum view of the world is here to stay, and I truly believe more people should become acquainted with this theory and everything it entails. From quantum entanglement and

16

Quantum Physics for Beginners

how it might explain things like destiny, to the very fabric of chemistry and solid matter, quantum physics is, I'm afraid, at the core of pretty much anything one can imagine.

Our power to advance computing, for example, lies in how quantum physicists will be able to solve equations from hereon. The Artificial Intelligence-fueled future of our science-fiction movies and books is much closer than we think, and, perhaps unsurprisingly, at this point, quantum computing lies at the very basis of this evolution.

Everything we are, everything we've ever been, and even the fabric of our Universe— they are intrinsically connected to quantum mechanics. Although it is a theory that seemed wild and unlikely a little over one hundred years ago, it has now become a *de facto* in the world of science.

With its mishaps and its strangeness, quantum physics is the very fuel we need to propel ourselves towards the shiny, beautiful future we have always wanted. Time travel, bending space, the ability to fully understand the Big Bang, who we are, and where we are heading—these are the wonderful, hard-to-answer questions quantum physics might make more straightforward for us.

Quantum Physics for Beginners

We have a long way to go between now and that future, it is true. But I believe that if we all focus our interest in building this future the best way we can, it will be far more within our reach than most people think.

I cannot urge you to become a quantum physicist tomorrow. What I can recommend you do, however, is to seek knowledge over a cookie-cutter recipe, to wonder and ask questions, to read and watch documentaries, and to try and learn as much as you can about the marvelous, stunning mysteries of the Universe.

You have a world of books and the Internet at your disposal, and I hope the book you are now holding in your hands will be the beginning (or even a continuation) of a beautiful journey into knowledge, science, and beauty.

Thank you again for choosing to walk the intricate paths of quantum mechanics with me. I appreciate it, and I truly hope the pages ahead will not disappoint you!

Chapter 1: Waves and Particles

To an untrained eye, the world of science and physics might seem pretty quiet. You wouldn't think that scandals of all sorts roam in the realm of science, or that scientists have very complicated relationships with each other, just like tabloid celebrities do.

For the vast majority of the history of physics, this has been the case. Nobody argued with Archimedes when he discovered the law of buoyancy, and Newton had no rival when an apple fell on his head (and a new era in physics began as a result of that).

Sure, scientific quarrels have existed and will always exist, but the world of physics has never been so split (yet so fantastically *cohesive*) as it is today.

The reason? Quantum physics.

Somewhat completely torn away from traditional physics yet somehow intrinsically connected to it, quantum mechanics (as quantum physics is sometimes referred to) is at a borderline between scandal magazines and science fiction movies.

Quantum Physics for Beginners

After nearly a century since its birth, quantum physics still raises eyebrows, crinkles foreheads, and makes scientists wonder. Then again, such is the beauty of science in general—it keeps on delivering answers that raise even more questions.

Waves and particles are at the very core of the chasm between traditional and quantum mechanics. Even if you hated physics in school, you could still imagine why they are very different.

A wave moves, it wiggles, it's everywhere and nowhere at the same time. You cannot hold it, touch it, or move it around in any way. Yet, it's well, *there*.

A particle, however, is something very much tangible. You can hold it, touch it, and move it around. You can toss it at your cat to play with it. It's *there,* and you know it.

Traditional physics viewed light as a wave. Things were simple (well, not *simple,* but they did not get as intricate as they are today). Everyone accepted that light behaves as a wave, and peace ruled over the blue skies of the scientific world. There were different directions in this theory, but overall, everyone agreed that light is a wave, and matter is made out of particles.

Quantum Physics for Beginners

Until, sometime at the beginning of the 20th century, a bunch of guys thought it would be a good idea to re-think this whole "light waves" thing. What they came up with might sound mind-boggling, but I promise that it is easier to understand than it sounds.

Namely, they started to discuss the idea that all particles or quantum entities can behave as both waves and particles. If you thought nothing in the world could be two things at once, welcome quantum physics into your life, and you will change your opinion.

Max Planck was among the first to start doubting what was, by then, the status-quo. While working on his blackbody radiation observations, he noted that to explain what he was noticing, he needed to consider light as "chunks."

Later on, Albert Einstein himself came up with a similar theory, saying that light is made out of little packets of energy (which he called "photons").

All of a sudden, light was not just a wave anymore, but a particle, too. As it frequently happens, this led to even more questions in the science world. In 1920, Louis de Broglie proposed that since light has energy, momentum, and wavelength, matter might follow the same pattern, and it might behave as a wave too.

Quantum Physics for Beginners

The relationship between waves and particles was baptized as "the wave-particle duality," and the most fun part about it is that big boy scientists are still arguing over it. To be honest, it is interesting to follow up on the topic, so, by the end of the book, you will be convinced that quantum physics is absolutely awesome.

To date, scientists have given a series of interpretations to the theory of the wave-particle duality and how matter can behave as waves and particles. Some of the most popular ones include:

The Copenhagen Interpretation

According to this theory, matter "drops" from its wave state when it is observed. Although this interpretation is the most widespread one, and makes a lot of sense, it is usually dismissed on the ground of, well, not actually being proven. Experiments have shown that this theory might be *the one*, but when it comes to explaining the mechanism behind it, scientists still have not found an answer.

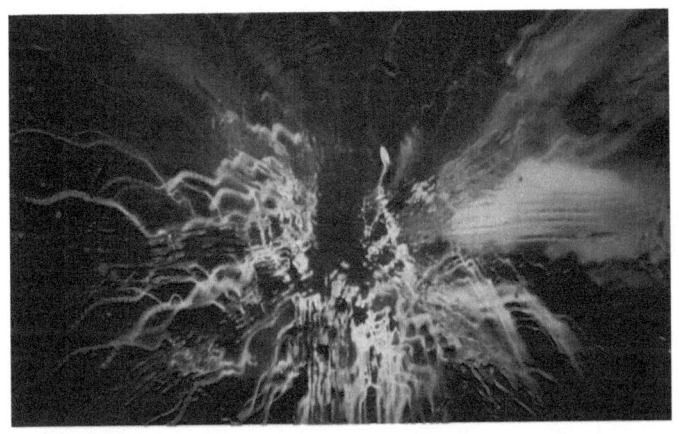

The Hugh Everett Many-Worlds Interpretation

Before we dive into the explanation of this interpretation, we have to make a quick pit stop along the way and introduce you to a theory that is usually considered to be more complicated: The superposition principle.

What this principle says is that when two (or even more waves) of the same type cross each other, the displacement of the point in which they intersect is equal to the sum of movements connected to each individual wave.

In plain English, if two waves overlap, the result of their disturbance is equal to the result of the disturbance each of them would have individually.

The Hugh Everett Many-Worlds Interpretation of the wave-particle duality states that the only existing wave function is the superposition of the Universe. In this view, the wave function will never collapse, so there's no need to measure its "fall."

The issue with this interpretation is that the act of measurement in it is connected to the interaction between the quantum entities (e.g., between the observer and the measuring instrument).

The De Broglie-Bohm Interpretation

In this interpretation, the information describing a system comes with a wave function and a trajectory that sets the position of the particles. In other words, the wave function creates a velocity field for particles.

As such, when something interacts with the system during the measurement procedure, the waves will separate in the configuration space (which makes them appear like they are "dropping" into particles).

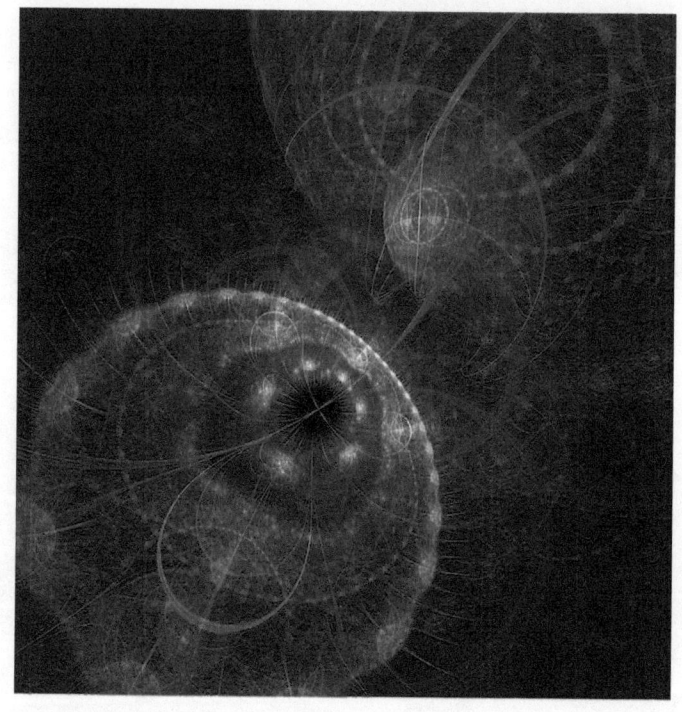

The Ghirardi-Rimini-Weber Interpretation

According to the Ghirardi-Rimini-Weber Interpretation, the wave function (the "drop") happens spontaneously. This happens because

Quantum Physics for Beginners

particles have a non-zero probability of going through wave function collapse.

In this interpretation, the wave function collapse is extremely rare (it happens every hundred million years). However, the number of particles in a system will most likely determine whether or not a collapse will occur in that system. In other words, the collapse of a single particle in a system will bring with it the collapse of the entire measurement apparatus.

The Quantum Decoherence Interpretation

The Quantum Decoherence Interpretation was created in the 80s. This theory does not describe the wave function collapse but how quantum probabilities become ordinary classical possibilities.

If anything, the wave-particle duality taught us all that answers are not always easy, and do not always come with a clear A or B answer. When someone asks whether electrons, for example, are particles or waves, no scientist in the world can ever give a clear answer:

Quantum Physics for Beginners

They are neither; they are both. They behave like both particles and waves, depending on a series of mysterious yet fascinating circumstances.

Does this duality have any real-life applications?

Well—yes and no. The wave-particle duality is mostly used in electron microscopy to allow scientists to view objects that are a lot smaller than anything that they can see in plain light.

This might not pose that great an interest to you, but the application might lie at the basis of a new invention to propel us into a whole new era. More importantly, though, the wave-particle duality made us question what we thought we knew, and shifted our thinking into a new paradigm. This is one that, as you will see, might be the very beginning of the science-fiction future we've all been fantasizing about. Teleportation, explaining destiny, finally finding out where we come from and where we're going—the answers might lie within the questions that quantum physics brings forward.

Isn't this fascinating?

Chapter 2: Identical Particles and Atoms

Identical particles and atoms are an interesting topic in quantum physics. Although it might be a little more challenging to understand, please bear with me. Just as in the case of the wave-particle duality, the basic concept behind identical particles is quite simple.

In classical physics, identical particles are distinguishable, which means that, although they are the same in terms of structure and the elements that comprise them, they can be distinguished from each other because there are marks that make them different.

In quantum physics, that goes out the window—not entirely.

What quantum physicists have discovered is that, when two particles are identical, they can change place with each other, and you will not be able to distinguish between them. This means that we could take two different particles from the Universe, swap them over, and nothing would happen. As you will see later on, this can have tremendous implications for our everyday life and how we might evolve from hereon.

Furthermore, it is worth noting that any kind of particles can be identical, from the simplest to the most complex ones (such as hydrogen atoms, for example). This can have immense implications as well.

Pauli's Exclusion Principle

Quantum Physics for Beginners

Pauli's Exclusion Principle is another important note we need to make when talking about identical particles. According to this principle, as elaborated by Wolfgang Pauli (Libre Texts, 2020), the electrons in an atom or molecule cannot have the same four electronic quantum numbers. One of them has to be in an up-spin, while the other one has to be in a down-spin. Particles that have an integer spin, like bosons, for example, are excluded from this rule.

Pauli's Exclusion Principle manages to bring about a deeper understanding of a series of matters:

1. It explains why electrons stack on top of each other instead of occupying the same space.
2. It helped shift our knowledge on solid state. For instance, in conductors and semiconductors, molecular orbitals form a band structure of energy levels. In metals (which are strong conductors), electrons reach a level where they are so degenerate that they cannot contribute to the thermal capacity of the material. This specific implication of Pauli's principle has been connected to mechanical, electrical, magnetic, optical, and chemical advances in technology.
3. The Uncertainty Principle of Heisenberg has also been connected to Pauli's discovery. This

principle states that you cannot know everything about particles at the same time.

4. In astrophysics, white dwarf and neutron stars are the demonstrations of Pauli's principle. These stars are collapsing, but due to a series of circumstances, they have a very high mass. Neutron degeneracy, the application of Pauli's Exclusion Principle in astrophysics, says that two electrons cannot occupy identical states even under the heavy conditions of a dwarf or neutron star. Because the neutrons do not overlap, they are filled with energy and forced to "accept" higher amounts of energy. This creates pressure within the star and prevents it from collapsing even further.

Identical particles might not sound like the most exciting topic under the sun. However, they represent a fundamental (and, at this point, rather unshakeable) chapter of quantum physics, so we couldn't have skipped them from our discussion.

Chapter 3: About Neutrinos

If you vaguely remember your physics class in school, you probably also remember that you were told that everything in the Universe is composed of atoms. Furthermore, atoms themselves are composed of protons and neutrons.

That was it—life was simple, and everyone agreed on it.

Until quantum physics came along and smart guys discovered that the world was divided into much smaller units (which is, in fact, the entire scope of the study of quantum mechanics in general).

Neutrinos are among these small units. They are among the smallest of the smallest units. Even more than that, they are utterly mysterious in the way they function and the way they can be "caught" and analyzed.

The first reference to the existence of neutrinos came in 1930 from Wolfgang Pauli (they were not called like that back then, though). It wasn't until 1955, however, that the first neutrinos were actually "captured" and observed, and ever since then, they have fed us and the science world with more questions than answers.

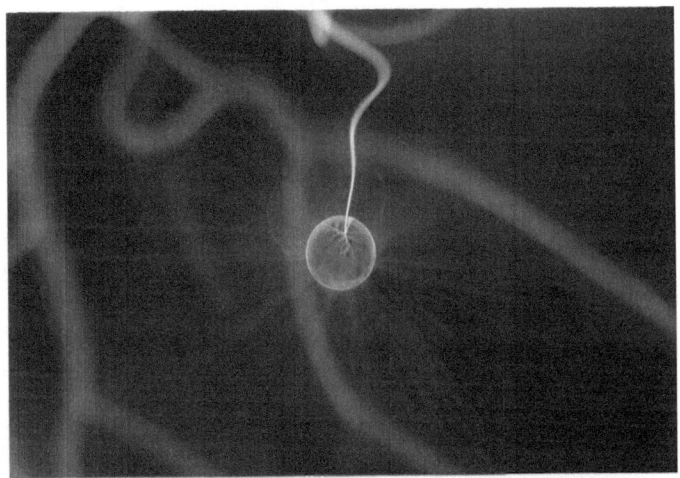

Neutrino Oscillation

There are a lot of mysteries in the field of neutrinos, but one of the most important and confusing ones is the question of neutrino oscillation.

We have already established that neutrinos are incredibly small particles. They are about one million times smaller than an electron, for example (just so that you can get a broad picture of just how tiny these guys are).

One of the more fascinating aspects about them is not their size, though, but that they have a very ghost-like behavior. Namely, they carry information across tremendous lengths of space without ever coming into contact with anything else, unaffected by magnetic fields, which makes them peculiar and of great interest at the same time, too.

Neutrinos are everywhere. They are around you just as you read this book; they move through you; they are in the sunlight and out there, in the big Universe. You will never actually see one, though.

Where do they come from?

Well, pretty much everywhere. They travel through billions of years carrying information about the places from whence they came, but they can be borne in pretty much everything in this world—including your body. For example, the radioactive decay of potassium in your body will produce neutrinos.

Likewise, cosmic rays that hit atoms in our planet's atmosphere can too produce neutrinos. They can also be generated in particle accelerators and nuclear reactors, for example. However, the neutrinos carrying the highest amounts of energy are borne far out in space, where something (yet undetected by humans) produces massive amounts of radioactive decay.

One very fascinating mystery about neutrinos is called "the solar neutrino problem." Namely, it is related to the fact that, whenever scientists attempt to observe neutrinos, they have always noticed fewer of them than the calculations showed. Now, you might believe that the estimates were wrong, but it is highly unlikely that an entire community of scientists didn't do the math right.

The explanation they have found, though, is stranger than you may even imagine. Aside from being ghostly particles, neutrinos are also tricksters, and they might change their typology.

To understand how this works, you should first understand (at least at a high level) that leptons and antileptons generate elementary particle reactions. A "lepton number" is the difference between the two, and neutrinos come in three main lepton flavors: Electron, muon, and tau.

When scientists calculate how much they can observe an electron coming from the sun's rays, they consider it as an electron. However, as it seems, neutrinos can change their "lepton flavor" on their path to us, meaning that they might become taus or muons. This makes scientists' jobs harder, as taus and muons are a lot more challenging to measure due to their lower energy levels.

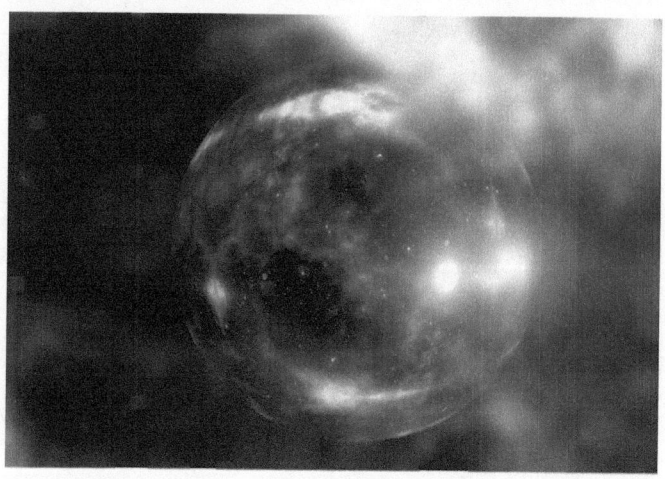

Why Neutrinos Matter So Much

Even though they are everywhere and in massive numbers, capturing them is extremely difficult precisely because they rarely interact with anything in the Universe.

Quantum Physics for Beginners

Scientists are making great efforts to capture neutrinos, and they only manage to do it every once in a while. Sometimes, when they do, they even name them because they are so rare and precious.

Capturing neutrinos is of great value to science people, though, as these little particles carry with them a lot of valuable information. One way to do it is by putting transparent materials in their way and waiting for the neutrinos to interact with the nucleus of an atom to reveal itself.

For example, in Antarctica, a group of scientists attempted to capture neutrinos at a facility called, very suggestively, *IceCube*. Here, scientists used a one cubic kilometer block of millennia-old ice that was extremely transparent for the capturing of neutrinos in more than 5,000 detectors shot through the ice cube.

Imagine billions of neutrinos traveling through an ice cube with thousands of detectors inserted in them. It is, indeed, an image larger than life. And yet, although you might expect that scientists have already studied at least thousands of neutrinos by now, you'd be shocked to learn that only about ten neutrinos are captured every year. Even more, only a handful of them are strong enough to give scientists enough information.

Quantum Physics for Beginners

When they are detected, neutrinos could provide us all with information about the farthest points of the Universe. They could tell us about those undetected sources of amazing energy, and they might even influence how we perceive the whole story of Big Bang. They might also help us gain a deeper understanding of black holes and everything they represent.

Neutrinos might be small, indeed, but they are a quintessential part of everything around us, and they deserve all the attention they are getting right now.

In many ways, neutrino astrophysics is the next step following infrared, gamma-ray, and x-ray analysis of the world around us. If we manage to understand these little guys and where they come from, we might be able to answer some of the most stringent questions we have ever asked ourselves as a species.

And we *will*, eventually, manage to do all that. Thanks to quantum mechanics, questions that we didn't know how to "grab" before are now a matter of technicality. The more we perfect our processes (such as the process of capturing neutrinos), the closer we will get to answers, and, who knows, perhaps new and exciting questions as well.

Quantum Physics for Beginners

Chapter 4: Laws of Quantum Physics

The laws of quantum physics were developed nearly a century ago to explain atomic and subatomic phenomena. At first, many physicists were skeptical about the new findings. However, they soon managed to pull all the information together to create a set of laws that govern quantum mechanics (or wave mechanics, as it is also sometimes referred to).

What you should understand, more than anything, is that the laws of quantum physics show that the impossible can be, in fact, very much possible. There is still a long way to go in this field until everyone accepts it and until we see actual practical applications on everything, but even so, things have settled a lot in the last couple of decades.

As a Nobel laureate in quantum physics, Richard Feynman famously said that, "If you think you understand quantum mechanics, you don't" (Ball, 2013). Many people interpret this as a sign that they should not even try to gain an understanding of this field, but the truth is that what he meant is that quantum physics frequently leads to both answers and questions (as we have touched upon in previous chapters as well).

Quantum Physics for Beginners

The truth of the matter is that we (as a species) *do* understand more of the fundamental laws that govern quantum mechanics than most even imagine. We might have a long way to go in deepening them and finding solutions to problems that don't allow us to push this theory further into even more practical applications. However, most physicists believe it is only a matter of time until we move past these hurdles as well.

The laws that dominate the small particles that comprise up everything are strange, as they are frequently very different from the rules that govern large objects. However, they are not entirely impossible to understand, as Feynman's quote might suggest.

Mind you, though: Although quantum mechanics laws govern the very small world, they also tell us a lot about actual reality and, as mentioned before, they might be able to tell us a lot about the grander scheme of things and the biggest and most complicated questions humankind has ever asked itself.

Let us not get ahead of ourselves, though.

Quantum Physics for Beginners

Why, exactly, are the laws of quantum mechanics so strange, and yet so important *and* fascinating at the same time?

Well, here are some things that are completely odd when you get to the functioning of subatomic levels in pretty much everything that exists:

- Things are in more than one place at a time (kind of), as particles do not like being "tied down"; they exist everywhere at the same time.
- What one particle does in one place can have an immediate consequence somewhere else, even if there is no one (or nothing) there to inflict this change.
- Particles can be everywhere until you look at them.

All of this is quite weird, right?

Even though scientists cannot always explain how these things happen, they have managed to test every one of these postulates in hundreds, maybe thousands of tests. As strange as these laws might seem, they are always proven right by experiments so much so that it is impossible not to believe the impossible that these laws represent.

Quantum Physics for Beginners

Let's dig a little deeper into some of the weirdest fundamental laws of quantum physics:

Quantum Leaps

One of the strangest things about quantum mechanics (or, better said, how particles function at a microscopic level) is the fact that particles can jump from their trajectory in ways that seem completely random.

Quantum Physics for Beginners

Imagine Earth is spinning around the Sun on its usual orbit, and then all of a sudden jumps on, let's say, Venus' orbit. Just like that.

Niels Bohr believed electrons could do this when they heat up. More specifically, he believed that when electrons heat up due to the energy they generate, they jump to another state. As it has always happened in quantum mechanics, experiments proved that he was right—even though Albert Einstein didn't quite agree with it (we will discuss his opinion and opposition to quantum physics laws later on in the book).

More recently, in 2009, it was discovered that although quantum leaps *do* exist, they take a little bit of time and are not as instantaneous as the scientific community believed until now (Minev, Mundhada, Shankar, 2019).

Even more, quantum leaps can now be controlled as well. Niels Bohr would have only dreamed of this, but recent research has shown that we do have the power to control these not-so-mysterious-anymore leaps (Crane, 2019).

This could have tremendous implications for quantum computing, for example. If you didn't know, quantum computers are computers that can process information a lot faster than regular computers because they can recognize the existence of zeroes and ones at the same time. Although there might be some errors to fix in quantum computers, the fact that we can now control quantum leaps means that we might soon be able to make better use of this new, exciting, and ultra-powerful type of computing.

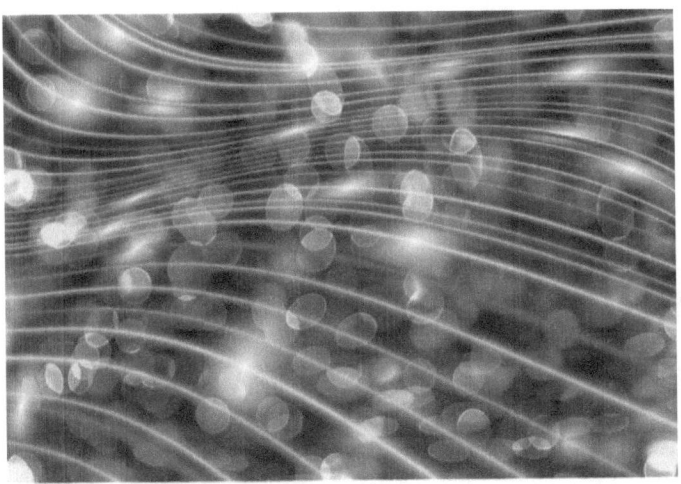

Quantum Entanglement

If you thought quantum leaps are weird, let me introduce you to quantum entanglement—the phenomenon according to which certain particles are intrinsically connected even when they are split by vast distances.

Think of it this way: Twins report that sometimes, they can experience the same emotions even when they are very far away from each other and not speaking. Likewise, entangled particles can too follow a given path, both when they are near each other and when they are, perhaps, thousands of miles apart.

The implications of this theory (proven by experiments) are mind-boggling. Quantum entanglement might somewhat explain the idea of destiny, for example, which—let's face it—is a very odd thing to be defined by science.

In more practical ways, quantum entanglement could help us finally *beam up* just like they did on Star Trek. If we manage to measure and control this phenomenon, we might be able to teleport ourselves (but we will talk a little more about this in chapter eight).

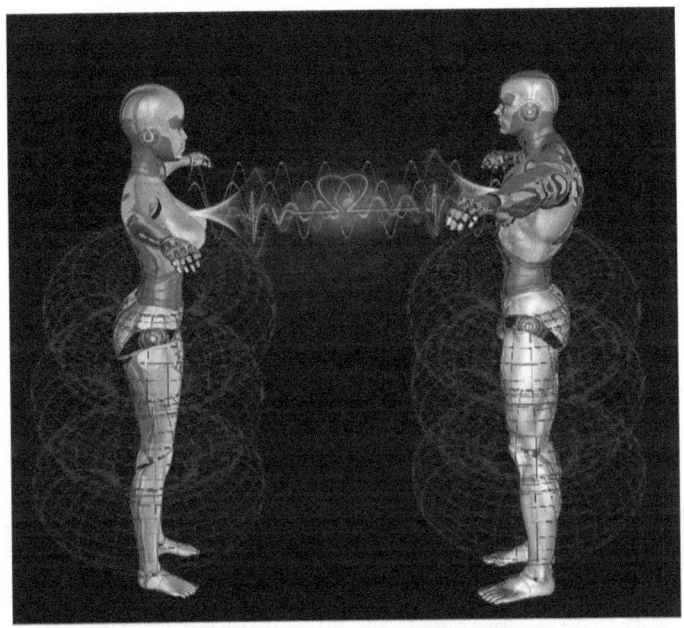

Quantum Spin

Quantum spin is a very odd thing to explain—even actual physicists sometimes have a hard time trying to put the entire phenomenon in plain words. That's how the term "quantum spin" was born.

Unlike what the name might suggest, a quantum spin has little to do with actual spinning. The phenomenon refers to an intrinsic angular momentum noticed in fundamental particles. The action itself makes particles seem like tiny spinning balls. The keyword here is "seem," because particles do not spin (the way the Earth revolves around the Sun, to bring this to the *macro* scale of things).

This so-called spin has two values: Spin-up and spin-down, which you can reach if you measure the angular momentum of the electron along a given direction. Unlike large objects that spin, the electron spin can never have "zero" as its state. Fundamental particles will *always* spin along an axis (whichever axis you might think of).

Quantum spin lies at the foundation of many discoveries—some of which have helped us advance our technology, others of which have helped us understand everything around us better. Along with quantum leaps, entanglement, and quantum fields, spins are among the most important elements to learn about when it comes to quantum mechanics (and pretty much all of the adjacent fields as well).

Quantum Physics for Beginners

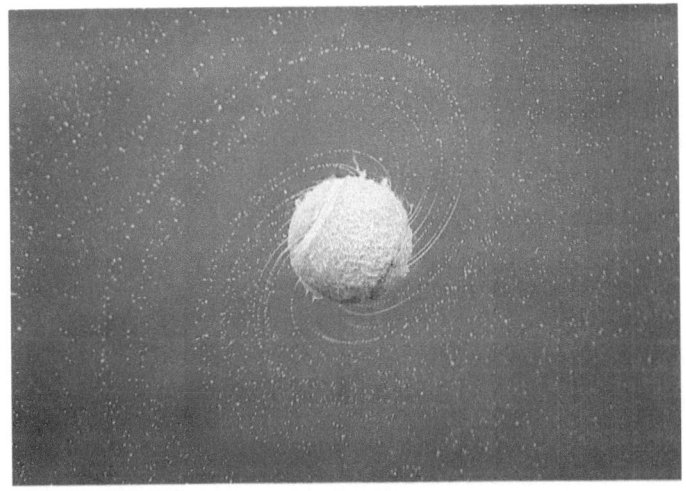

Things Are Unpredictable

A general rule of quantum physics strangeness is that everything in this field is rather unpredictable. Gone are the days of deterministic physics of Sir Isaac Newton, and enter a world that we are barely beginning to understand.

Albert Einstein and many others had trouble with the uncertainty of the microcosmos around us. For the most part, they believed that the strange things particles do when we are not "looking" at them were just an error in judgment on our end.

Regardless of who you are and where you come from, the weird laws of quantum physics affect you—not necessarily in a bad sense, but in the sense that, whether you like it or not, they are there.

Even more, these laws will most likely shape our future, and it is a future far more exciting and fantastic than it may seem.

Chapter 5: The Uncertainty Principle

Despite its name, the uncertainty principle has led us to precise, almost palpable answers like why the sun shines, for example, or how the emptiness of space might not be that empty anyway.

Physicists started to think about the uncertainty principle towards the end of the 1920s, and it came from Niels Bohr's institute in Copenhagen, where a German physicist called Werner Heisenberg was studying the implications of the then-very-fresh quantum theory.

As mentioned in the previous chapters, the "original" quantum theory said that light could be described as both waves and little sachets called "quanta." On the surface, this might seem like complete gibberish to non-specialists; it doesn't change the taste of pizza, and it doesn't influence the things that make us happy like a loved one's embrace, right?

And yet, quantum theory has changed everything. Most of us do not even realize how many modern tools have roots in that simple theory which include things like LED screens, modern computers, or even thermonuclear energy.

The uncertainty principle was one of the first theories to be derived out of that first wave-particle duality. In essence, what this principle says is that there are things we cannot understand about how particles work. What we can hope for, however, is to measure and predict, with a certain amount of certainty, what these particles will do, and the repercussions their actions will have.

In more scientific terms, the uncertainty principle says that we cannot measure with absolute accuracy both the position and the momentum of a particle. If we manage to be very accurate in our measurements of the position of a particle, it means we will be extremely far away in accuracy when it comes to measuring the momentum of the same particle. The vice-versa is valid, too: If we are accurate in measuring the momentum of a particle, we will not be accurate about measuring its position.

The uncertainty principle explains, for example, why electrons orbit the nucleus of an atom and do not collide with it. By the logic of classical physics, the negative-charged electron and the positive-charged nucleus should, theoretically, attract each other, just

Quantum Physics for Beginners

as a positively-charged magnet and a negatively-charged one would.

They don't, however, and this is something that has been admitted by traditional physics too. It wasn't until the uncertainty principle that scientists were finally able to explain it. Namely, what happens is that if the electron were to get closer to the nucleus, we would be able to measure its position with extreme accuracy. However, we would not be able to measure its momentum (and velocity), and at some point, the electron would spin out of the atom altogether. That does not happen.

The uncertainty principle did sound like a lot of fun for physicists, but not all of them agreed with it. Among them was Albert Einstein himself, who, if we have to be honest, was skeptical about this whole quantum physics "thing" altogether.

To be more specific, Albert Einstein was perceived, for a long time, as a strong opponent of quantum theory (and the uncertainty principle by extension). One of the main arguments he had against the uncertainty principle is called "Einstein's box," a thought experiment he proposed to Niels Bohr.

The experiment proposed considering an "ideal box" lined with mirrors. This box would be able to contain light for an indefinite amount of time. Einstein also said this box could be weighed, and once that happens, a clockwork mechanism would open a shutter at a precise moment and release one photon, which means that the time at which the photon releases would be known. If the box is weighed after the release of the photon, it means that one could also measure the energy of the emitted light. In other words, the location, time, momentum, and energy levels can be measured with accuracy.

His argument was not wrong, but Bohr counteracted it, saying that the box would have to be weighed by a spring and a pointer. Since it would have to move vertically when its weight changes, there would be a level of uncertainty regarding vertical velocity. As such, it would show uncertainty in how high above the table the box is located. Even more, Bohr pushed this forward and said the elevation of the box concerning the Earth's surface would be uncertain too, precisely because, as Einstein himself had proven it, gravity affects time.

This was not the last "feud" Einstein had with quantum mechanics. His following one, the Einstein-Podolsky-Rosen paradox (EPR paradox), gave birth to a concept we have already discussed: Quantum entanglement. Through this, Albert Einstein remained inscribed as one of the leaders of quantum physics. However, it is not entirely clear how *certain* he was of the entire theory, especially since it slightly goes against what he had proven before.

Why Is the Uncertainty Principle Important?

As with all theoretical physics, the uncertainty principle can feel a little *uncertain* when it comes to applications and how it influences day-to-day life, especially for people who do not necessarily have much tangency with science in general.

Quantum Physics for Beginners

The uncertainty principle is important because it lies at the basis of alpha decay—a type of nuclear radiation. Most often, heavy nuclei such as uranium-238 are connected to two protons and two neutrons which are bound inside the nucleus. Usually, they would need a lot of energy to escape this. However, because an alpha particle has a well-defined velocity, its position is not very well-defined. There is a small chance the particle could move out of the nucleus at some point, even if it doesn't have the energy to escape. When this happens, the process becomes "quantum tunneling" (the particle has to somewhat dig its way out of the nucleus, and it cannot leap).

Similarly, you can observe quantum tunneling in the center of the sun. Here, protons fuse to release the energy that makes our sun shine. Normally, the temperature at the center of the sun is not high enough to allow protons to gain a lot of energy (not enough for them to overcome electric repulsion, at least). However, the uncertainty principle "helps" them tunnel their way out of the sun's core and through the barrier of energy.

There are many other explanations of the world that relate to the uncertainty principle, including vacuums. Before quantum physics, scientists explained vacuums as the absence of everything. Within quantum

Quantum Physics for Beginners

physics, however, and due to the uncertainty principle, they can sometimes represent the temporary absence of energy from a quantum system. This might have significant implications on how we understand space, for example, as well as the "vacuums" within it.

Chapter 6: Quantum Field Theory

When, in the introduction, I promised you that this book would bring forward some of the most pressing age-old questions of humankind, I was not joking. That is what physics in general deals with, and what quantum physics, in particular, has managed to *somehow* find answers to.

Let us not get ahead of ourselves here, though. It is important for your understanding's sake to discuss this chapter step by step, methodically, to make sure that all the information you already know ties into the new bits that will be presented here. I promise it will be worth it because quantum fields might be considered as the *building blocks* of the entire Universe.

People have asked themselves *what we are made of* for a very, very long while now. We have found religious and spiritual explanations; we have written poetry and music about this, and, to an extent, classical physics has found pertinent answers, too.

Quantum Physics for Beginners

Never before have we gotten so close—so within reach of the ultimate answer, though.

In our more traditional view of the world, everything that existed was made out of particles. More specifically, everything was made out of the chemical elements included in the Periodic Table of Elements—and that was it. Again, things were simple and relatively easy—they had a deterministic force to them that made people feel at ease that "the science guys are on it" with whatever happened.

Until, *boom*, quantum physics happened, and things were suddenly not that well-known, or set in stone, as before. This is actually a good thing because, while it is useful in a vast range of situations, the Periodic Table did not provide us with too many actual answers when it comes to the origin of everything and how it is that everything exists (ourselves included).

What quantum field theory says, in essence, and with a very short explanation (because I don't want to make you feel too confused), is that we should consider the field upon which the particles we study to be quantum as well.

What is a "field" in this understanding?

Quantum Physics for Beginners

Well, a field is "something" that stretches across the entire Universe. It takes specific values at different points in space and allows the same values to change in time. In some ways, the "quantum field" is a fluid creating ripples throughout the Universe.

Basically, before quantum theory, scientists could not find all the answers they needed when it came to how the subatomic levels function. Part of this was related to the fact that, although they had agreed that the Universe is uncertain, they had not changed their vision of the field in which their studied particles are being studied.

It's like knowing you are gluten-insensitive, not changing the recipe for the pizza dough, and then wondering why you are still getting the same results.

The moment physicists started to consider fields as quantum (and thus, governed by the same laws of quantum mechanics), they managed to gain a much better understanding of processes like:

- Radioactive decay (described in the previous chapter)
- Matter creation and annihilation
- Quantum corrections to be made to the electron magnetic moment

Quantum Physics for Beginners

For the entire field of quantum mechanics, the quantum field theory was the missing link everyone (including Einstein) had been looking for—the bridge that would finally start to connect quantum physics and classical physics.

Why Is Quantum Field Theory Important?

Although quantum theory was a pretty massive storm for the world of classical physics, the vast majority of theories developed by "old school" physicists still stand. Gravitation is still here, for example—it's just that quantum theory (and quantum field theory) might help us understand it better.

Quantum field theory also allowed scientists to create the Standard Model, a sort of "Table of Elements" adapted to the world of quantum physics, which describes how different particles interact with each other. Even more, the same model predicted the existence of other yet undiscovered particles.

Unlike the Periodic Table, which included no less than 118 elements, the Standard Model explains everything with six quarks and six leptons. In addition to this, the Standard Model predicted the existence of the infamous Higgs boson, a particle that was believed to offer other particles their mass. Also known as "God's particle," this boson was first actually discovered a few years ago at CERN, as a particle accelerator.

In fact, you might remember their experiment, when back in 2012, CERN scientists managed to isolate the boson. When that happened, mainstream media was somehow "activated" against it for fear that it would open a black hole.

Alas, no black hole was opened (at least none we know of), but Stephen Hawking (who was entirely skeptical about the possibility to isolate the Higgs boson) famously said that although this boson might be credited to have generated the Big Bang, it might also be the one to wipe out the entire Universe one day (Dickerson, 2014).

Spooky, right?

It might be, but as I keep on emphasizing throughout this book, this might also be the beginning of a whole new era in our evolution as a species. Surely, most of the new discoveries in science and technology were a little spooky at first. I bet people were terrified of planes, for example, when first launched, and that they didn't understand computers very much either. I bet cars were a weird thing then, and I bet the first steam engine was pretty scary as well.

We have all come to get accustomed to all these "novelties," and they are now part of our lives in such a way that we could never even imagine life *without* them. Perhaps theories like the quantum field one *will*, one day, be as "normalized" as, let's say, Pitagora's theory in mathematics.

We cannot do anything but wait and see what comes next.

Chapter 7: Schrodinger's Theory

Of all the names associated with quantum mechanics, Erwin Schrodinger's is one that most people have heard of (at least remotely). There are many jokes on the internet about his theories and you might have stumbled upon some of them. Yet, let me tell you this: What he said is not a joke, but is vital information about how we perceive reality.

In this chapter, we will talk more about Schrodinger and the cat, who is almost as famous in the world of physics as Pavlov's dog. Don't worry, though, this thought experiment has nothing to do with pets, but has everything to do with quantum mechanics.

Schrodinger's Cat

No cat has ever been as famous in the scientific community as Schrodinger's. While Pavlov's dog was used to exemplify the power of habit, Schrodinger's cat theory was a little fuzzier (and frankly, a lot more abstract).

To understand Schrodinger's thought experiment, we must first define a couple of terms, one of which is "superposition" which is also known as "quantum superposition." This is the physical state that includes all potentialities. Remember how we said that, in quantum physics, a particle could be everywhere? Well, this state of the particle being everywhere (and nowhere in particular) is called "superposition."

In some theories, the particle assumes a specific location and state only when the particle undergoes a "wave function collapse." Erwin Schrodinger believed that this collapse only happens when the particle is observed.

To exemplify his theory, he proposed the following thought experiment: You have a box with a cat and a vial of radioactive substance in it. After one hour in the box, one of the atoms in the radioactive substance has an equal probability of decaying and *not* decaying at the same time. There is an equal probability that the cat will either be alive or dead in the box. The only way one can find out if the cat is alive is by opening the box, which means an external observer makes the quantum superposition collapse into one of the possibilities or the other. Until then, the cat is both alive and dead in the box.

A lot of scientists thought the theory was ridiculous, precisely because it goes against everything we all deem as "normal." How could a cat be both alive and dead?

Quantum Physics for Beginners

Erwin Schrodinger's cat experiment was not about showing that the cat is alive and dead at the same time, though. It was about exemplifying how quantum superposition and wave function collapse happens and just how important it can be, too.

Before his thought experiment, he famously enunciated an equation meant to complement Newton's laws of motion (applied to the large scale matter) with a series of laws that would be correlated to the small-scale matter. The equation is known as the "Schrodinger's Equation," and it describes the form probability waves take in their governing of the small-particle motion. The same equation also specifies how external influences can alter these waves.

His findings are now used in atomic, nuclear, and solid-state physics. However, his cat thought experiment has faced a lot of criticism in its day, and even today, scientists are still debating the idea.

Schrodinger's theory is known as just one of the many interpretations of quantum theory. There are others you should probably get familiar with because they will help you gain a deeper understanding of this entire field (and the fascination surrounding it).

Quantum Physics for Beginners

How wave functions collapse and become one of the many possibilities is a question that has been "bothering" physicists ever since quantum theory first came out. So far, no clear answer has been given, but there are a few theories that might be able to explain how this happens.

Most physicists ascribe to the Copenhagen interpretation (as described in the first chapter of this book), which is where Schrodinger positions himself as well. However, scientists are still considering whether or not this theory makes sense, especially since there might be situations in the Universe where it would be more than just odd for it to be true.

For example, black holes absorb everything in their way. However, according to Schrodinger and the Copenhagen school of thought, the black hole wouldn't be able to do this without an observer. Most scientists are not entirely satisfied with this idea, so they have come up with a series of "tweaks" to Schrodinger's theory to eliminate the observer out of the equation and "allow" wave function collapse to happen independently. These versions of the theory are usually ascribed under the "objective collapse models" range.

We might not have clear answers just yet, but you have to admit, quantum physics has helped us understand a lot about the granular functioning of the entire Universe. Even more, it might yet have a lot of interesting things to unveil, like the topic we will discuss in the next chapter, for example.

Chapter 8: Quantum Teleportation

Of all the crazy science fiction tools humankind has ever thought of, teleportation is by far at the top of the list. Back in the 1970s, in the movie *Star Trek*, people were beaming on and off of ships, yet most of those who watched the show never thought the idea would be palpable in real life.

And yet, here we are. Thanks to quantum mechanics, teleportation is not that much of a crazy thought anymore.

Remember when we discussed quantum entanglement? As we were mentioning it back in the previous chapters, this theory might be the foundation of teleportation. Believe it or not, the thought of transporting ourselves from New York to Paris in a matter of seconds is not *just* pure science fiction anymore. It is something actual physicists are actively working on.

There is still a pretty long path to walk before that, but never before in the history of science have we ever felt that this dream was within our reach. It might never come true, but it might, as well, become palpable reality.

How Quantum Entanglement Relates to Teleportation

The concept behind quantum teleportation is relatively easy to understand if you already know how quantum entanglement works. What the theory says here is that the information of an atom or photon could be transmitted in its exact state from one place to another.

At a larger scale, this would mean that objects and humans could be transported from one location to another via means of quantum entanglement too. For instance, you could enter a cabin in New York where all your particles are scanned and later rearranged in another cabin in Paris. The spookiest part about this is that the "original" you would probably vanish in the process of "scanning." Therefore, the whole quantum entanglement concept better be absolutely certain that it can reconstruct you at your destination—or else, things could get really really dark, soon.

Of course, we are to trust scientists on this one. Before any such device would be put to use, numerous tests would have to be run to ensure that it works and that it does not pose any risk. As mentioned earlier, *flying* was pretty risky in its inception days as well, yet, here we are, flying from one location on Earth to another without a second thought.

How close are we to quantum teleportation, exactly?

Well, we have known that quantum teleportation is theoretically possible ever since the 1990s. However,

Quantum Physics for Beginners

in 2019, scientists actually managed to achieve the unimaginable: Teleporting a photon, which, by the way, is considered to be a three-dimensional state (Haughton, 2017). Before that, in 2017, scientists had also managed to teleport a photon from Earth to its orbit, which was in a two-dimensional state (Spender, 2017).

A photon is pretty far from an actual human being (or even that Amazon parcel you've been waiting on), and two years seems like a lot of time to move from transferring two-dimensional objects to transferring three-dimensional ones. However, the more experiments are successful, the more matters will accelerate in this field, so some individuals do hope to witness human teleportation within our lifetime.

Until then, however, scientists have also shown that we can transmit information between two computers via means of quantum entanglement, which might seem completely useless considering we have the Internet at our disposal. But it might also be the beginning of a new era in communications (Nield, 2019).

While most scientists are wary of predicting just how far quantum entanglement could go (i.e., if we will ever be able to "beam out," for example), one thing is for certain: Quantum teleportation *is* possible. It is but a matter of "scalability" to see if objects that are more massive and more complex than a simple particle could be seamlessly transferred from one place to another.

These are matters only the future can tell.

Chapter 9: The Zeeman Effect

Electromagnetism pretty much controls everything in the Universe. You cannot see it nor touch it, and yet it is there, keeping everything together. Without electromagnetism, our planet would spin out of its orbit and wander into the cold emptiness of space (and guess what would happen to humankind next?).

The Zeeman effect is one of the experiments that reminds us of how important electromagnetism is. This effect was discovered towards the end of the 19th century by Peter Zeeman, a Dutch physicist who received the Nobel prize for the same discovery in 1902.

It is more than worth mentioning that Zeeman's discovery came at a time when quantum mechanics and quantum theory were not even a materialized thought. Even so, most people ascribe the Zeeman effect to be at the confluence between classical physics and quantum physics, and you will soon see why.

What Peter Zeeman discovered is that spectral lines will split if they are in the presence of a magnetic field. Without quantum mechanics, nobody knew how to explain his discovery, so it took several decades before a theory was formulated.

The Zeeman effect is sometimes categorized in two typologies: The normal Zeeman effect and the anomalous Zeeman effect. This effect appears because the electron orbitals become distorted in the presence of a magnetic field. Without a proper understanding of spin described by quantum mechanics, the Zeeman effect was understood classically. However, when quantum mechanics came along, and scientists discovered that certain electrons have spin, they also started to understand the Zeeman effect through the new perspective as well (and as such, the "anomalous" Zeeman effect was born).

What happens during the Zeeman effect is that an electron interacts with a magnetic field. This makes the electron have two magnetic moments: One that is associated with orbital motion, and one that is related to the spin.

Today, the Zeeman effect is used in a variety of fields, including but not limited to astrophysics. For instance, scientists are using this effect to determine how the "solar spots" function (these are spots on the Sun where the temperature is lower due to the magnetic field rays). This helps astrophysics specialists understand how the magnetic field of the

Sun works, for example, and it might one day be extrapolated to how other stars in other solar systems work as well.

Furthermore, scientists have used the Zeeman effect in determining the energy levels of atoms and how they identify with angular momenta. Last but not least, the Zeeman effect is a good way of studying atomic nuclei and electron paramagnetic resonance.

The Zeeman effect is frequently correlated with the Stark effect, which is similar, but instead of showing how a magnetic field affects nuclei, it shows how an electric field affects them.

In the grander scheme of things, and the issues that have been presented in this book so far, the Zeeman effect shows that there *can* and *should* be a reconciliation between classical physics and quantum mechanics. As you will see in the following chapter, things were not always considered to be as such, and Albert Einstein was right in the middle of this "scandal" (I promised, the world of quantum physics can be quite scandalous in itself).

Chapter 10: The No-Cloning Theorem

The no-cloning theorem is one of the most interesting theories in quantum mechanics even if you look at it just through the prism of its strangeness (because, yes, quantum physics does not disappoint when it comes to this).

Out here, in the "real world," we always copy things. We copy and paste text from one document to another, we copy art, and we might even try to copy the way of being of someone else. And it's acceptable (within the laws of copyright, of course, and within the laws of common sense as well).

In the quantum world, copying is impossible, and not because there's a "quantum police" that will not allow you to copy particles, but because such are the weird laws of quantum physics.

In short, the no-cloning theorem states that an arbitrary unknown quantum state cannot be identically copied. It is worth mentioning that this theory is valid for pure states and not for mixed states, and that it is a direct consequence of the phenomenon of superposition.

What Are the Implications of the No-Cloning Theorem?

The no-cloning theorem doesn't exist just because some crazy scientist thought of adding a new rule into the world. As mentioned above, this rule exists because it is a direct consequence of superposition. Cloning or copying particles in their quantum states cannot be done because the copy would theoretically "come out" in superposition.

Quantum mechanics is linear, so if, for example, we were to copy simple arbitrary states, we would, in theory at least, get the results we are expecting. If we were to copy complex linear constructions, however, we would go against the linearity of quantum mechanics because the act of copying would, in itself, be a non-linear action.

We cannot make perfect copies of a quantum state because, due to superposition, we do not know how to select the right "cloner" for the right state. In other words, cloning (or copying) information to absolute perfection would contradict Heisenberg's uncertainty principle because it would automatically mean that each variable is very precisely measured.

The no-cloning theorem (and its practical proof) might seem utterly annoying at first. However, it has been the foundation of quantum cryptography precisely because it automatically prevents any "third parties" from creating copies of the information and stealing them.

How about teleportation—how does that relate to the no-cloning theorem, then?

This is a pretty iffy topic among physicists these days, but teleportation might not be entirely possible precisely due to this theorem. The one way that is palpable and realistic is by ensuring that the first copy of the teleported object is completely "cut" (instead of "copied") and then pasted at its new destination.

In essence, this means that, if we were to teleport as humans, the original "us" would completely disappear when "beaming out," and a new, identical version of us would be reconstructed at a given destination. As was mentioned in chapter eight, though, this idea is spooky, so it might not be feasible for humans or other living beings to teleport.

They might, however, be able to teleport your Amazon parcel, which, let's admit it, is still kind of cool, right?

Chapter 11: Bell's Theorem

Bell's theorem is one of the most commonly misunderstood theories in all of science (not just quantum mechanics). What it does is provide an answer to the EPR paradox (which we will discuss in this chapter as well).

Bell's theorem bears massive importance in quantum physics because it proves that the early formulations of quantum theory were right and that the opposing school of thought, with Einstein as one of its leaders, was incorrect in their assessment.

The EPR Paradox

The EPR paradox is one of the most famous arguments against the early versions of quantum theory. Developed by Einstein, Podolosky, and Rosen (hence the name), this theory postulates that the earlier previous of the quantum theory go against the theory of relativity as Albert Einstein formulated it.

Quantum Physics for Beginners

To help you understand this argument, and why Bell's theorem finally shed a more conclusive light on it, we will go back a little to a topic we have already explained: Quantum entanglement.

The EPR paradox is a thought experiment that starts, like all quantum entanglement examples, with two particles. These particles are entangled with each other according to the rules we have already discussed in chapter four. According to the Copenhagen school of thought, each of these particles is in an uncertain state until it is measured, so it is everywhere and nowhere at the same time until someone or something measures it.

When one particle becomes certain through measurement, the other one becomes certain as well, which leads to a paradox because, through entanglement, they communicate with each other at speeds faster than the speed of light. This comes in contradiction to Einstein's theory of relativity.

When this thought experiment was first proposed, things were a little fuzzy to most scientists (on both ends of the argument). No matter how much they argued and debated on it, they did not know how to expand and be more precise about the example.

Quantum Physics for Beginners

Later on, however, David Bohm tweaked the EPR paradox a little to make it clearer. He postulated that if you take an unstable spin 0 particle that decays into two different particles (A and B), which head in opposite directions, the sum of the two particle spins is 0 (because the initial spin was 0 as well). As such, if particle A has a spin of $+\frac{1}{2}$, then particle B must have a spin of $-\frac{1}{2}$ (and the other way around).

According to the Copenhagen school of thought, neither of the particles would have a definite state until a measurement occurs. They are in a superposition of all their possible states and show an equal probability of having a positive or negative spin. However, as soon as one of them is measured, both of them will assume the right state, which means information between them travels at speeds faster than light, and reverts us to the idea that Einstein's theory of relativity is in contradiction to the rules of quantum entanglement.

Bell's Theorem

Perhaps odd to a non-specialist, Einstein and his fellows never argued with the second part of the thought experiment, as it was a logical consequence of the "rules" set by the first part that entangled particles communicate with each other.

Einstein did not have an issue with entanglement as such (which, by the way, he called "spooky action at a distance"). What he thought, however, was that quantum theory was still incomplete and that there are missing variables scientists did not yet know of.

Spoiler alert: He was wrong. Even after hundreds of experiments and theoretical formulations, Einstein's so-called "spooky action at a distance" happens just as the original quantum physicists thought. And yes, it still contradicts the general theory of relativity.

Although not a theorem per se, but a collection of results coming from an experiment that measured what happens when particles communicate faster than light and what happens when they can't do that, Bell's theorem was one of the stepping stones in finally putting the debate to rest.

Quantum Physics for Beginners

When John Stewart Bell devised this experiment, many people thought that it would finally be able to prove that particles were not communicating with each other at speeds faster than light. By extension, this would have meant that they weren't doing weird things while we were not observing them, either.

To many people's surprise, however, Bell's experiment proved that particles do communicate at speeds faster than light. However, scientists concluded that, just because this one instance of quantum mechanics is correct, it does not mean that the entire theory is correct.

Perhaps even more importantly, it also proved that even if quantum theory is correct, it does not automatically rule out the theory of general relativity. Let's say, for example, that all cats are black. This does not automatically infer that all animals are black, so the feud between Einstein's theory of relativity and quantum theory is not necessarily based on a logical consequence of thoughts.

The general theory of relativity and quantum theory is still not fully reconciled. However, it is a work in progress physicists today are working on, so we might be able to see the results of their experiments during our lifetime.

Why Is Bell's Theorem Important

In essence, Bell's theorem is important because it shows us that sometimes, things can appear to go against each other, but that this does not necessarily mean that they cannot survive, more or less harmoniously, in the same world.

Classical physics and quantum mechanics are still torn apart from each other, precisely because what stands true in the first is what the latter does upside down. Where things in the "big world" are ordered and logical, they are also extremely messy and uncertain in the granular levels of everything that surround us.

It does not make much sense, but it is what it is for the moment, and all we can do is wait and see what scientists come up with from hereon. It will be fascinating to watch, though, even if just as an amateur bystander.

Chapter 12: Einstein's Coefficients

As I hinted throughout the book, Albert Einstein did not have a very steady and smooth relationship with quantum mechanics. He didn't hate it, and he most definitely did not dismiss it altogether, especially given the fact that he was one of the first people to admit that the then-new quantum theory was about to change the rules of physics as a whole.

Before we jump to the core topic of this chapter, Einstein's coefficients, we will discuss a little about his relationship with quantum mechanics. This is necessary because it will help you better understand where Einstein's coefficients theory falls within the grander scheme of things.

Einstein vs. Quantum Mechanics

Einstein's "affair" with quantum mechanics is frequently misunderstood, as a lot of people place them against the current, thinking that he did not agree with quantum theory as a whole.

That is not entirely true, however. Einstein believed everything could be measured and predicted. However, as mentioned before in this book, quantum mechanics is not that measurable mostly because when the measurement act happens, wave function collapse occurs as well.

Unlike Niels Bohr and the Copenhagen school, Einstein believed this is mostly related to our inability to understand the phenomenon, rather than some mysterious ways of the Universe. Everything a lot of quantum physicists find borderline-magical about this field, Einstein found to be just a question without an answer (yet).

In many ways, his view of quantum mechanics was quite healthy; it was what most people in their full mental capacity would think. To understand why his opinion was the voice of what seemed reasonable back then, you have to know, first and foremost, that quantum physics represented a turning point for the entire field.

Things were deterministic until then. Everything had a cause and an effect. Everything could be measured and predicted with a high level of certainty.

"God does not play dice with the Universe," Einstein famously said, referring to the fact that quantum uncertainty did not pertain to the natural order of things (Dickerson, 2015).

Quantum Physics for Beginners

Quantum uncertainty came and tore everything apart, and it did not even go directly against any of the laws or theories of physics enunciated before it. As shown by the evolution of our understanding of the Zeeman effect (presented in chapter nine), quantum mechanics explanations were long due. It was only a matter of time until someone (or a group of people) was to figure that out.

Albert Einstein did not like the uncertainty surrounding quantum theory. He felt a better explanation was on its way, and the uncertainties were a matter of definition and points of view, rather than a reality of the Universe.

Furthermore, Einstein might have taken quantum theory, and specifically, the uncertainty principle, very much personally as well. If the uncertainty principle were to be right, he believed the theory of general relativity that made him famous was incomplete and potentially wrong as well.

These days, the gap between Einstein's relativity that dealt with the very large scale of things, and Heisenberg's uncertainty principle is getting narrower by the day, as physicists are working on ways to bring the two together and finally unify classical and quantum mechanics.

Quantum Physics for Beginners

Going back to the feud between Einstein and Bohr, one would expect that the former won over the latter, somehow. Albert Einstein is, after all, the one physicist everyone has at least heard of, right?

Wrong. As it seems, there were a series of debates between Einstein and Bohr, and the latter won in the end. Einstein proposed arguments against the Copenhagen school as it was represented by Niels Bohr, Heisenberg, and others, but his arguments were dismissed. He then attempted to propose the thought exercise we have already talked about (Einstein's box), but Bohr managed to dismiss that as well.

By far and large, Einstein was most likely wrong. Time and again, quantum physics has been proven by experiments. One after the other, they all confirmed that quantum theory is, in fact, real, albeit, *a little* strange, as we have discussed before as well.

Does This Dismiss Einstein?

Alright, so Albert Einstein was wrong about quantum theory—or, at least, up to this date, nobody has been able to prove him right.

However, that does not dismiss his many other accomplishments. The general relativity theory is, perhaps, one of the most famous formulae in history (we all know *E equals MC square*, right?). To date, his theory has numerous applications, and modern physicists are trying to reconcile it with quantum theory.

In addition to this, Einstein enunciated a series of other theories that are still used in practice today. The Einstein coefficients—the core topic of this chapter, are among them.

In essence, Einstein's coefficients are mathematical quantities meant to measure how probable light can be absorbed or emitted by an atom or molecule. There are two coefficients, according to Einstein:

- A coefficients, which relate to the rate of spontaneous light emission
- B coefficients, which relate to the absorption/ stimulated emission of light

Also, according to Albert Einstein, both of these emissions are needed to achieve Planck's radiation law. For a bit of background, this law had been devised in 1900 by Max Planck, a German physicist, to explain the spectral-energy distribution of radiation emitted by a black body (Encyclopedia Britannica, 2020).

There have been numerous applications of this theory, including how a semiconductor laser should be operated (Taylor, 1990). Einstein's coefficients are largely considered to have contributed to the invention of lasers as a whole.

Up until today, Albert Einstein remains very much relevant. It might seem like this book, and the entire realm of quantum mechanics is off against him. However, the truth is that he pertains to a world that is not fully reconciled with that of quantum physics. He dealt with the rules of what happens at the largest scales of the Universe. Quantum mechanics, on the other hand, deals with very small scales. Between the view of quantum physics and that of classical physics, there is a connection physicists are working on. When that equation is solved, we will likely uncover some of the biggest secrets of the world around us.

Don't take this book as an attempt to put classical physics down in any way. Rather, take it as a way to broaden your knowledge and understand the intricacies of science, if not in-depth, then at least at a higher and wider level.

Conclusion

Without a doubt, quantum mechanics *is* essential. The nature of everything that surrounds us lies in complicated mathematical formulae elaborated by tens, maybe hundreds of brilliant physicists' minds and several decades of debates, calculations, and questions we have yet to answer.

Take solid matter, for example. To most of us, the existence of solid matter is not even something to question. After all, you are here, reading this book on some solid device, and you are quite solid, I must assume.

For physicists, this is not that simple a question.

In fact, it took them no less than four decades to formulate, mathematically, how it is that solid matter exists (from a quantum point of view). I will not get into the gory details of how they proved this, but they did it in 1967.

Without quantum mechanics, solid matter would never have been proven by an equation. It sounds absurd, but scientists need equations and formulae for everything, precisely because, like it or not, it is equations and formulae that lie at the very advancement of humankind as a whole.

What else wouldn't have existed without quantum theory?

Well, chemistry. The periodic table of elements was devised long before quantum mechanics came into the world. However, it was this branch of physics that helped us gain a deeper understanding of each element in this table and how complex molecules come to exist.

Why is quantum physics so essential for our understanding of organic chemistry?

Well, you see, chemistry happens because the electrons in each element on Mendeleev's table behave a certain way. Without quantum physics, we would have no idea what that behavior is, and, as such, we'd have no idea how to work with different elements either.

Quantum theory lies at the very basis of creation, and understanding it means that we might finally be able to push the frontiers of our current knowledge. Knowing where we came from, how the Big Bang came to be, how we can move past the limits of our current technology and explore other solar systems (and maybe, *just maybe* other galaxies, too)— these are questions we've always asked ourselves one way or another.

Ancient Greeks and Norsemen found spiritual explanations for the world and everything that was going on in it. The stars, the thunderstorms, the way the rain fell on summer evenings, whether or not the sun would shine enough light to grow crops—these were questions mythology has attempted to answer.

Even more, following mythology, philosophers and mathematicians have also attempted to answer the same questions. And for a good chunk of time, even with the limited resources they had back then, they managed to paint a surprisingly accurate view of the world (except, perhaps, that which said the world is flat or that which said Earth is at the core of the solar system).

Physics, as a discipline, was derived from the observations of the ancient Greeks, for example. The very word was drawn from the language they spoke, actually: *Phusis* meant "nature," and as such, physics became "the study of nature."

For centuries, we did just fine with knowing the world is in perfect order. Some argued that this order is the "hand" of a supernatural being or force that

Quantum Physics for Beginners

took care of everything, like a Marie Kondo of the Cosmos, ensuring every little thing is in its place. Deterministic science suited us just fine, but it had its limitations.

In the 20th century, everything broke into a million pieces, and not just in physics. In literature, poets were writing poems that did not rhyme anymore, and whose very format was weird as if to mirror the brokenness of the 20th century.

In the arts, cubism and other modernist currents were taking shape, also to show that the perception we had over ourselves was shattered into pieces. Look at any painting signed by Picasso at his peak, and you will understand.

Nietzche declared that God is dead, the old world was slowly taking its way out and leaving room for a new, industrialized, heavily-machined world where news could travel faster than ever, and so could death.

In physics, the large, deterministic view of the world we had gotten so accustomed to broke into tiny little pieces we now call "particles", and for the last hundred years (roughly), reconciling the old, and the new in physics is what we have been mostly preoccupied with.

Quantum Physics for Beginners

The nuclear bombs the US dropped over Hiroshima and Nagasaki at the end of the Second World War came as a sort of morbid, terrible crowning of the first half of the century. Science seemed to have prevailed, but behind closed doors, humankind was still wondering why.

Indeed, my dear reader, quantum mechanics can lead to terrible, atrocious discoveries. But it can also help us understand the nature of our reality at a scale we never hoped to do before.

Even more, quantum theory might be the *push* we need to step into a new world—you know, the kind that has been portrayed by *Star Wars*, *Star Trek*, and *Dune*. Their future seems so much more technologically advanced than ours that it almost feels impossible to reach out and *think* it would become a reality. And yet, there's a good chance it will become our day-to-day life.

Sooner or later, the "theory of everything" (as scientists now call the theory meant to bring together classical physics and quantum mechanics) *will* come to be. Every year, physicists are getting one formula closer to finally knocking it off and finding the truth about how we function as humans, as planets, and as specks of dust in the massive Universe.

Quantum Physics for Beginners

I hope this book has helped you understand that not only is quantum physics important, but that it is not something of which you should be afraid. Indeed, things can get intricate and complex the more you advance in this field, but the absolute truth is that they can get exponentially more fascinating as well.

From the fundamental rules of quantum theory to applications and the famous feud between Einstein and "his guys," quantum mechanics has gone a long way. Indeed, it will most likely have a long way to go yet, but the road we have traveled so far is promising.

There is a future of wonder and amazement waiting for you right around the corner— a future in which the Universe itself is shaped like our wildest dreams and where nothing is unreachable.

We might be wrong, but there is a very big chance that quantum physics will get you there. You may never get so deep into this field as to learn its complexities, but even just a fundamental knowledge (that I presented in this book), can go a very long way.

All the experiments made until now have proven that quantum mechanics is not, in any way, a dead end. Given that all theories are proven true by actual practice, it is, hopefully, only a matter of time until we

Quantum Physics for Beginners

finally uncover the "secret key" that has been hiding from us.

Truth be told, the issue of reconciling quantum mechanics and classical physics under a so-called "theory of everything" might be the longest-standing and most interesting problem physicists ever had. For decades now, they have been ruminating over the same questions. Indeed, they do not have a complete answer just yet, but little by little, the puzzle is coming together.

What awaits us around the corner? Infinite possibilities, but more than anything, the most in-depth understanding of the Universe we'll ever have.

Thank you for choosing to walk this path with me in *Quantum Physics for Beginners Who Flunked Math and Science*. Thank you for coming this far, and for bearing with me.

More than anything, thank you for being curious. The future belongs to those like you, who wonder and question, read and acquire information, and who are passionate about moving forward. I wish you the best of luck for your future!

Bluesource And Friends

This book is brought to you by Bluesource And Friends, a happy book publishing company.

Our motto is **"Happiness Within Pages"**

We promise to deliver amazing value to readers with our books.

We also appreciate honest book reviews from our readers.

Connect with us on our Facebook page www.facebook.com/bluesourceandfriends and stay tuned to our latest book promotions and free giveaways.

References

Ball, P. (2013, January 24). *Will we ever... understand quantum theory?* BBC Future. https://www.bbc.com/future/article/20130124-will-we-ever-get-quantum-theory

Crane, L. (2019, June 3). *Quantum leaps are real – and now we can control them.* New Scientist. https://www.newscientist.com/article/2205089-quantum-leaps-are-real-and-now-we-can-control-them/

Dickerson, K. (2014, September 8). *Stephen Hawking Says 'God Particle' Could Wipe Out the Universe.* Live Science. https://www.livescience.com/47737-stephen-hawking-higgs-boson-universe-doomsday.html

Dickerson, K. (2015, November 19). *One of Einstein's most famous quotes is often completely misinterpreted.* Business Insider. https://www.businessinsider.com/god-does-not-play-dice-quote-meaning-2015-11

Haughton, R. (2017, July 20). *Quantum teleportation is even weirder than you think.*https://www.nature.com/news/quantum-teleportation-is-even-weirder-than-you-think-1.22321

Libre Texts. (2020, May 18). *Pauli Exclusion Principle.* Chemistry LibreTexts. https://chem.libretexts.org/Bookshelves/Physical_and_Theoretical_Chemistry_Textbook_Maps/Supplemental_Modules_(Physical_and_Theoretical_Chemistry)/Electronic_Structure_of_Atoms_and_Molecules/Electronic_Configurations/Pauli_Exclusion_Principle

Minev, Z., Mundhada, S., Shankar, S., & et al. (2019). To catch and reverse a quantum jump mid-flight. *Nature*, 570, 200-204.

Nield, D. (2019, December 31). *Physicists Just Achieved the First-Ever Quantum Teleportation Between Computer Chips.* Science Alert.

https://www.sciencealert.com/scientists-manage-quantum-teleportation-between-computer-chips-for-the-first-time

Planck's radiation law | Definition, Formula, & Facts. (n.d.). Encyclopedia Britannica. https://www.britannica.com/science/Plancks-radiation-law

Sagan, C. (1994). *A Pale Blue Dot.* Random House.

Spender, T. (2017, July 14). *Is teleportation now a reality?* BBC https://www.bbc.com/news/science-environment-40594387

Taylor, G. (1990). The use of Einstein's coefficients to predict the theory of operation of a semiconductor laser. *Journal of Applied Physics, 68*(7), 3122-3139. doi: 10.1063/1.346407